Why Engineers Need to Grow a Long Tail

Why Engineers Need to Grow a Long Tail

A primer on using new media to inform the public and to create the next generation of innovative engineers

---

Bill Hammack

You may download a pdf of the complete book at http://www.engineerguy.com

Publication Data

Hammack, William S., 1961-

Why Engineers Need to Grow a Long Tail: A primer on using new media to inform the public and to create the next generation of innovative engineers

p. cm.

Includes Further Reading and About the Author

ISBN 978-0-61539-555-5

1. Engineering -- United States. 2. Engineering -- Social Aspects -- United States. 3. Engineers -- Public Opinion -- United States. 4. Technology -- Social Aspects -- United States.

Contents

Preface **v**

Introduction **1**

The New Media Landscape & Its Effect on Science Journalism **3**

New Media Isn't Just Old Media Delivered in a Different Way **13**

How Engineers Should Talk to the Public **21**

New Media in Action: Two Hypothetical Case Studies **33**

Conclusion: Creating Zing! **45**

Further Reading **48**

About the Author **51**

Preface

This very short book - a pamphlet really - began as a series of lectures given over a ten-year period. It's based on talks given at the National Press Club, to the Council for Chemical Research, to various universities and national labs, and in informal speeches in the corners of dining rooms when I spoke to small groups. I thought of expanding it to be a "big" book but instead chose to distill and concentrate the ideas to the essence of what an engineer needs to know. My hope is that practicing engineers, who are often busy, will use it as a quick start guide. It boils down my own observations and practice as a person who truly works in the trenches of public outreach: every day I'm in a studio or an editing suite preparing some media piece to share with the public - a piece that defines what an engineer is, what he or she does, and why their work is important.

Bill Hammack
Urbana, Illinois

1
Introduction

OFTEN the details of new media get lost in an alphabet soup that usually begins with an "i" - the iPod, the iPad, the iTouch - but new media also includes URLS, YouTube, and so on. Yet the essence of new media is not *in* these devices, but in their *use.* Thus I intend this short book to be a primer on how to *think* about new media.

I emphasize "think" because this book isn't about technique. You won't learn best practices for Facebook or Twitter. Surely by the time the ink even dries on this page a new generation of devices and web applets will be available. So, I focus on the deeper issues of communicating in a user-generated era. My hope is for engineers to design rich new media entities that revolutionize how they communicate with the public and that attract the next generation of engineers.

To do this, engineers need to grasp the mindset of new media and to understand the underlying changes in the media landscape that will outlast the latest social networking tools. I carefully separate two related but indcpendent questions. First, what *exactly* is new media? Second, regardless of the medium, *what* should engineers communicate to the public? Thus, chapters two and three serve as a quick guide to understanding the Web 2.0

landscape; the fourth chapter addresses at length what engineers should be telling the public. The "what" doesn't depend on whether we tweet in the blog-o-sphere or make good old-fashioned television shows - although eventually the impact may differ a great deal between the two approaches. Clearly, future growth lies in new media.

Thus, I hope this book encourages practicing engineers to develop new, powerful ways to reach the public and to help them understand what engineers do and why engineering is important. I hope that educators - both classroom teachers and engineers educating informally - use the ideas here to create the next generation of innovative engineers.

A generation ago, Marshall McLuhan famously said, "The medium is the message." Never has this been more true than today. While someone of my age may look at a YouTube video as a novelty, to a younger person it is *the* way to communicate. America's youngest generation expects to get their information from content-rich social media. As Clay Shirky has pointed out, new communication tools become socially interesting when they become technologically boring. For young people today, the new social tools are beyond normal: they are heading to ubiquitous and will likely be invisible soon.[1] The engineering profession needs to have a meaningful presence in Web 2.0 before invisibility fully arrives, otherwise our use of the medium will appear ham-handed and graceless.

With this book, I hope, perhaps immodestly, to prevent such an outcome.

---

[1]Shirky, Clay, *Here Comes Everybody: The Power of Organizing Without Organizations* (New York: Penguin Press, 2008).

2

The New Media Landscape & Its Effect on Science Journalism

IF YOU can *fully* answer the question "Why did Madonna drop her record label and replace it with a concert promotion company?" you can skip ahead to the next chapter - you completely understand that we live in a new media world. If not, remove your records from the turntable, turn off the VCR and give me a few minutes to share some facts and figures that will give you a way to think about the revolution occurring in the media world.

I recognized the need for such an overview when I mentioned to a colleague that I'd been on public radio's *Marketplace* the night before, and thus had reached about six million people. He said "That's all?" The degree to which you express incredulity at his answer might serve as a litmus test for reading this chapter about as well as any questions about Madonna's career.

Audience Numbers

A key aspect of my career has been the realization that engineers need to make mass media an integral - perhaps *the* integral - part of our outreach. We have many great programs that work at the local level – mobile units like my university's "physics van" which transports demonstrations to students - but what we really need is to dramatically leverage our time. That calls for mass media. This requires, of course, understanding the current environment so that we can see where to fit in. We should be able to figure out what size audience we can realistically aim for and to anticipate audience trends. So, let's take a look at audience numbers in order to get a feeling for the media landscape and our goals.

Fragmentation as measured by sitcom finale

I start with the "big three" networks' evening national news programs.[1] (Do you still watch this? I stopped in 1984!) ABC, NBC and CBS have about six to nine million viewers. That number alone isn't interesting, but over the last twenty-five years, network news has lost one million viewers each *year* -- that's half their audience in the last twenty years.[2] This fact reveals an essential truth about the expansion of the television dial and the fragmentation of the audience.

I've developed a new statistic to illustrate this splintering of the dial, which I call "fragmentation as measured by sitcom finale." Here are the relevant data:

Viewership for final episode

M*A*S*H (1983)	106,000,000[3]
Seinfeld (1998)	76,300,000[4]
Friends (2004)	51,100,000[5]

[1]For those under fifty: Up until the mid-1990s the dominant source of information for most Americans was the nightly network news on one of the only four networks ABC, NBC, CBS, and PBS. These broadcasts got huge audiences and drove the news cycle. The "king" of the news, if you will, was Walter Cronkite, who retired in 1981.

[2]http://www.journalism.org/node/943, State of the News Media 2008, Journalism.org.

[3]AP David Bauer, February 4, 2008, story on The Super Bowl.

[4]*New York Times,* March 16, 1998.

Since *M★A★S★H,* the final episodes of very popular shows have lost about 25 million viewers each decade or so. Is *Friends* 50% "worse"than *M★A★S★H*? No, it isn't that *Friends* is a lesser sitcom than *M★A★S★H* - I mean, neither of these is *Charles in Charge* - but rather the dial has fragmented. We can see this fragmentation clearly if we study ratings for various news and information media - or at least what passes for news and information today.

Listenership & viewership for today's news/information programs (March 2008)

Rush Limbaugh (radio)	13,700,000
Morning Edition (NPR)	13,200,000
Evening Network News	~7,500,000
O'Reilly (Fox News)	3,070,000
Dobbs (CNN)	1,222,700
New York Times	1,037,000
Hardball (MSNBC)	600,000

No doubt that by the time this is published some of shows listed might even be canceled, but the trends and punchlines are clear:

- Television has large numbers in the aggregate, but it has completely fragmented; often you are one of 100,000 or so watching a show on a cable network.
- Public radio has not fragmented and has gone gangbusters - this is an educated, voting, active audience whom we don't want to lose.[6]

[5]*Multichannel News,* August 14, 2006.

[6]Some facts about the demographics of the public radio audience: These listeners

- Printed newspapers are on the decline. The top 20 papers have lost about 10% or so in circulation in the last two years, and their circulation further drops every quarter. The printed newspaper lost its economic model when Craigslist took over the classified ads. Online newspapers are doing better, but there is no economic model to make as much revenue as the printed papers.[7]
- Engineering communicators need to look at economic ways to get chunks of 100,000 listeners and, when it can be done, a million or more.

With these numbers, I've put in perspective that reaching six million people with a commentary on *Marketplace* is pretty darn good! Still, the numbers reveal a story of crisis for journalism.

The crisis in journalism

Every day brings more dire news for journalism: declining circulation of newspapers, dropping viewership of television news, and fewer listeners for commercial radio. Nowhere is the impact more profound than on science, technology, engineering, and medical journalism. Ralph Cicerone, President of the National Academy of Sciences, clearly spelled out the crisis:

---

are intellectually curious and enjoy learning about the world around them. They are 33% more likely than the general population to express an interest in theories and 32% more likely to enjoy learning about art, culture, and history. This is an active audience. Over 70% voted in the most recent local, state, or federal election. NPR listeners are 22% more likely to be involved in clubs and organizations than the general population. NPR listeners are more than twice as likely to have addressed a public meeting, written to an elected official, or written to an editor of a magazine or newspaper. Approximately 9.3% of the NPR audience is African-American.

[7]Keep in mind the difference between circulation and readership. In this Internet age one can indeed have high readership of a newspaper website but low circulation of the printed paper. The problem is turning readership into income.

> *[W]e are also seeing troubling signs that communicating science, engineering, and medicine to the general public is getting harder. With recent downsizings at newspapers, magazines, and broadcast outlets, there are now fewer full-time science writers and less space or time for serious, in-depth reporting.*[8]

As print media retrenches, it often regards science journalism as a luxury. For example, in 2004 the *Dallas Morning News* let go of their esteemed, well-recognized, award-winning science editor in the wake of a costly circulation scandal.[9] A struggling *San Francisco Chronicle* laid off award-winning medical journalist Sabin Russell, who had covered health policy and medical science for twenty-two years at the publication.[10] The *Houston Chronicle* laid off its aerospace reporter of twenty years. The venerable *Boston Globe* got rid of its Health/Science section, moving health to the Arts & Lifestyle pages and relegating science to its Business columns. And in 2008, CNN completely dismantled its science, space, and technology unit.[11] According to Mooney and Kirshenbaum, only one minute out of every 300 on cable news is devoted to science and technology, or one-third of 1 percent. These changes are emblematic of a wider shift in viewer and reader habits that have affected the presentation of science on television.

---

[8]Cicerone, Ralph, "Celebrating and Rethinking Science Communication," *The National Academies InFocus,* Fall 2006, vol. 6, No. 3.

[9]Layton, Charles, "The Dallas Mourning News," *American Journalism Review,* April/May 2005.

[10]Mooney, Chris and Sheril Kirshenbaum, "Unpopular Science," *The Nation,* August 17, 2009.

[11]Brainard, Curtis, "CNN Cuts Entire Science, Tech Team," *Columbia Journalism Review*, December 4, 2008.

The fracturing of science television programming

Likely every reader over fifty recalls the great science shows of the 1970s: Carl Sagan's compelling *Cosmos* or Jacob Bronowski's majestic *The Ascent of Man*. Yet today the rise of cable and satellite has fractured the television dial into thousands of small pieces, each of which grabs a fraction of the audience of the past. The ratings for quality shows like PBS's *NOVA* have seen over a 50% decrease - typically, a *NOVA* episode sees an audience of one and one-half to two million viewers.[12] Other outlets for science programming, like the *Discovery Channel,* have fallen 30% in the last four years - from 1.3 million in prime time to about one million today.[13]

In addition to a declining audience, the fragmented dial has changed the quality and impact of television programing - impairing the ability to offer rich, detailed, and thoughtful coverage of science, technology, and health. Because the television dial features hundreds of channels, we have become a nation of "channel surfing" viewers. As a television producer once put it to me, "We don't worry that people will tune away, we know they will, we worry about bringing them back." This means that TV has become a land of sound bites and arresting visual images that may or may not have meaning. Images are chosen first and foremost for their ability to return viewers to the channel, not to convey meaning. The programmers of the *Discovery Channel*, for example, often make prime time represent a "theme" -

---

[12]Private Communication, 2006.

[13]Steinert-Threlkeld, Tom, "Dirty Work," *Multichannel News,* August 14, 2006, vol. 27, #32, pp. 18-20.

recent examples include "shark week" or a focus on "dirty jobs."[14]

We now live in a world of niche audiences

Is there hope? No, not in the sense that large audiences will ever be aggregated again. This issue is one that the journalism profession continues to struggle with, especially in looking for an economic model. The implications for our liberal democracy may well be profound: media choice might well increase inequality in political involvement and polarize elections.[15] Yet this troubled media landscape *does* offer an opportunity for the engineering profession.

In the age of monolithic audiences, which required expensive tools - networks, costly cameras, sophisticated microphones - engineers found it hard to be heard. It was difficult to get mentioned on the nightly news or to be featured in a television drama. Nobel Laureate Leon Lederman suggested development of a "television pilot that would instead show researchers as skeptical, creative romantics."[16] In the fragmented world of niche audiences, by using cheap digital tools and internet distribution, the engineering profession can now target and reach the

---

[14]As of this writing, *Discovery Channel* prime time consists of these "dirty job" shows: Garbage Pit Technician, Skull Cleaner, Geoduck Farmer, and Fuel Tank Cleaner. *Multichannel News* as cited above.

[15]See, for example, Prior, Makurs, *Post-Broadcast Democracy: How Media Choice Increases Inequality in Political Involvement and Polarizes Elections* (Cambridge: Cambridge University Press, 2007).

[16]*New York Times Magazine,* August 13, 1995, Section 6, page 16. Lederman worked with professional script writers, AAAS staffers, and even got funding from the National Science Foundation and the Department of Energy. He wanted to counter a growing anti-scientist feeling by presenting scientists with the same allure as the lawyers and doctors on *L.A. Law* and *ER*. He called it, "Scientists fall in love."

audience we want - audiences of perhaps the 100,000 I mentioned above. The power of these niche audiences lies in their engagement with what they've read or watched.

Internet users like to forward science & technology stories

Two University of Pennsylvania researchers studied how internet users share information.[17] Jonah Berger and Katherine Milkman learned that people preferred to forward articles with positive themes, and they liked to send long articles on intellectually challenging topics. "Science kept doing better than we expected," said Dr. Berger, a social psychologist and a professor of marketing at Penn's Wharton School. He continued:

> *We anticipated that people would share articles with practical information about health or gadgets, and they did, but they also sent articles about paleontology and cosmology. You'd see articles shooting up the list that were about the optics of deer vision.*[18]

That, of course, is exactly the type of engagement that engineers want for their message. To fulfill the potential of these niche audiences, though, we need to understand thoroughly how new media works and to understand how young people use media differently than their parents.

The younger generation has replaced Descartes' "I think, therefore I am" with "I have a webcam, therefore I am." No one under 25 uses e-mail any more; it's all instant messaging. Facebook now dominates in every campus computer cluster. The 20-something set even uses media communally: at

---

[17]Berger, Jonah and Katherine L. Milkman, *Social Transmission and Viral Culture* (unpublished research report, University of Pennsylvania, 2009).

[18]Tierney, John, "Will You Be E-Mailing This Column? It's Awesome," *New York Times,* February 9, 2010.

parties, five or six people might gather around a laptop and share their favorite YouTube videos. New media aren't just a different outlet, they fundamentally change how the media world works.

The concert is now king

For example, in the music industry the change from records to tapes to CDs was what I will call "linear." In other words, the sales model remained the same with each higher-resolution medium. The iPod, though, disrupted this chain. iTunes and the iPod have ended the age of the CD - music now arrives piecemeal, song-by-song, making little money. In 2000, record companies sold $13.5 billion worth of records. By 2008, this number had dropped to nearly half - $8 billion.[19] In fact, the very popular band Radiohead shocked the music industry by releasing their latest album for free. When they later released the CD in stores, however, it was the top-selling album! This phenomenon is part of the new rules that I discuss in the next chapter. The big payoff in music now lies in using the songs to bring fans in for large concerts. Madonna, for example, fired her record company and signed up to be managed by a concert promotion group. She signed a $100 million dollar contract with LiveNation, a company that specializes in concerts. The deal is simple: they give her $100 million, and she gives them records and the rights to license and sell merchandise. Clearly, in this day and age, LiveNation isn't counting on making back its money on record sales. Instead they hope to profit from ancillary streams, such as commercials that license the music, ticket sales, t-shirts, etc. Nothing illustrates the financial power of concerts and the arrival of a new media age more than the

---

19 EconTalk, "Meyer on the Music Industry and the Internet," March 22, 2010.

oldest rock band alive. In 1975, one could buy a concert ticket to see Mick Jagger strut for $8.50, which would be $34.00 in today's dollars. When the Stones toured in 2006, a ticket cost $100, a threefold increase in constant dollars.[20] Small wonder the Fox network organized a concert tour for the cast of *Glee.* One way to increase profits from their television show is to move their performers around the nation, instead of just broadcasting over the airwaves.

Isn't new media just a bunch of toys?

You're probably thinking that these new media are just toys, yet every new medium starts as a toy. The first copyrighted motion picture in the U.S. was *The Sneeze* by Thomas Edison. Second, we've been at these crossroads before, just with different media. In 1950, both television and 3D movies debuted. Many thought television to be a fad; some thought 3D movies were the wave of the future. That same holds true of the "new" media we have today. We don't fully understand this new landscape: some things will be duds, some will be fads, and some will become permanent parts of our culture. But if you think something like Facebook is a toy, keep this in mind: the *New York Times* and ABC News collaborated on a project using Facebook to deliver election news, including sponsorship of a debate.

The Facebook generation

So, the expectation of the Facebook generation is that they will be able to participate, create, and share multimedia. Science and engineering communicators need to participate in and even shape those media, both of which require a deep understanding of how and why new media works.

---

[20] *On the Media,* National Public Radio, October 23, 2009.

3

New Media Isn't Just Old Media Delivered in a Different Way

When I talk of "new media" or "Web 2.0" I don't mean simply delivering "old" media via the web. By "old media" I don't even mean a particular technology (movies, television, radio, newspapers, magazines, etc.) but instead a particular process. I've worked extensively in "old" media, so to illustrate that process let's look at the creation of one of my commentaries for public radio's popular *Marketplace*.

How old media works

Typically I pitch a piece to a sub-editor; we'll discuss the piece thoroughly, look for any holes, logical leaps, discuss the news hook for it, and also make a "snapper" for the ending. Together, then, we develop a script. That script goes to an editor or two above my sub-editor for approval. We then make changes, head to the studio, and carefully lay down the audio tracks - re-taping any parts that didn't sound just right. Usually we do the taping the day the piece airs, so a few hours after my studio visit the commentary appears on *Marketplace* and is then heard by six million people. Later, of course, it appears in a downloadable audio file - an MP3 - so it would seem this has a new media presence, yet it really doesn't.

What makes something "old" media is that process I

described of completely polishing a piece, filtering it through many gatekeepers (editors, sub-editors, and the like), carefully editing the final piece, and then offering it to the public. The essential characteristic of old media lies in this model: filter, then publish. The new media inverts this completely: one publishes and then filters. Think of a place - a repository or a searchable, browseable web space - where engineers place their videos reflecting their own interests and their take on engineering. Wouldn't this, just be a free for all -- a mishmash of video?

Isn't "publish, then filter" just a useless free-for-all?

One key to a successful "publish, then filter" site lies in adding a social dimension. If you look carefully at a site like YouTube, the public is able to rate and rank the videos. They do this astonishingly well: highly rated video are indeed interesting and sometimes informative. Or, consider a site like Flickr, which is designed to share photos. Flickr features two billion photos! One of the earliest Web 2.0 applications, it works as a photo repository fueled by social organization tools, which allow photos to be tagged and browsed by "folksonomic"[1] means.

For example, sixteen users pooled 1,712 images of Steuben County in upstate New York, including wineries and lakes, hunting and fishing, dining and shopping. No one person

[1]What a wonderful word! Here, from *Wikipedia*, is its definition: "Folksonomy (also known as collaborative tagging, social classification, social indexing, and social tagging) is the practice and method of collaboratively creating and managing tags to annotate and categorize content. Folksonomy describes the bottom-up classification systems that emerge from social tagging. In contrast to traditional subject indexing, metadata is generated not only by experts but also by creators and consumers of the content. Usually, freely chosen keywords are used instead of a controlled vocabulary. Folksonomy (from folk + taxonomy) is a user generated taxonomy."

set out to organize such a thing, no media outlet assigned a team to it, yet it does have value. Other members of Flickr sort and rate these photos, allowing a user to look only at the most interesting ones. On Flickr one can find thousands of these groups - the 219 members who took 2,271 photos of the latest Minnesota State Fair, or the 191 people who shared 5,719 images of the "Cans" Festival in London. So, while it seems that Flickr, Wiki, and YouTube have no quality control, in another sense they are completely quality controlled - many videos, wiki entries, or Flickr photos are never viewed, as they are deemed completely unworthy.

For example, someone started a Flickr group for the "British General Electric Company", which has only two members, one of whom contributed twenty-one of the thirty-three photos.[2] Even worse was the "LLI Liberty & Summit Conferences", which had one member who posted fifteen photos.[3] No one participated in these groups and they failed – just two of surely tens of thousands of such failures. So, failure in the "publish, then filter" world is high, but the cost of failure is low. What has changed in the last ten years - due to digital tools for video and sites for sharing with the world - is this dropping cost of failure.

Yet, even this doesn't fully explain the power of "publish, then filter." The descriptions above imply that the procedure works only to find the "hits" that appeal to a mass audience, and while this happens, it represents only half the power of

---

[2]For the curious: "This group is about the people, places and products associated with the GEC from its beginnings in 1886 until 1999 when it became Marconi plc."

[3]Also for those with an inquiring mind: LLI is "a group of entrepreneurs and students of personal development who are changing the financial and personal courses of our lives. As part of that journey we attend conferences all over the world in places like Melbourne, Rome, the Atlantis Resort (Bahamas) and Hawaii."

new media.

Beyond mega-hits: the long tail

The web has blurred the line between a private communication and a public broadcast. In the past one would never listen in on a phone call or open someone else's mail, and similarly one knows that a commentary broadcast on public radio's *Marketplace* is designed for all; yet, the web is filled with things like this:

> *A flower vendor was just packing up and he had a very nice, good sized rosemary plant. I was planning to cook a chicken tomorrow and missed the herb plants that I had at home, so I was glad to get a new one. On the way back to the tram stop, I stopped into Wilkinson's where at last I found a wastebasket.* [From a blog by Felicita written on September 27, 2008]

What is this? Surely something like this about a visit to the mall cannot replace the "old" media? It cannot, but implicit in this question is an error: Assuming those using new media are trying to find some common denominator to reach a mass audience as old media does. Or, more simply, put, "They aren't talking to you!" And we aren't really talking about audiences.

Social networking sites like MySpace and Facebook have millions of accounts, yet the median number of friends on MySpace is two, whereas the average is 55 - although the distribution isn't bell-shaped, it skewed toward lower numbers. This means that social networking is largely done pairwise: One person communicating with another. A blogger like Felicita is one of millions of pairwise (or perhaps tertiary or higher) interactions. So, from an "old" mass media viewpoint, an audience of tens or hundreds is a failure of sorts - yet audience is the wrong word to use. What Felicita has is a "community", a community in which she, for whatever reason, resonates. It is a secret of Web 2.0 (social networking) sites that one doesn't need professional quality in video, or narrative technique, or performance to be successful. The success of a content-rich site would be much like a dinner party: it isn't important what's on the plates, but instead what's on the seats. The social networking of Web 2.0 allows people to choose what appeals, rather than sit and receive coarse marketing messages, with a global communication cost so low the lowest common denominator in communication can be overcome. This means the tyranny of the most popular has been defeated by the long tail, a concept outlined in a popular 2006 book by Chris Anderson:

> *The theory of the Long Tail is that our culture and economy is increasingly shifting away from a focus on a relatively small number of 'hits' (mainstream products and markets) at the head of the demand curve and toward a huge number of niches in the tail. As the costs of production and distribution fall, especially on-line, there is now less need to lump products and consumers into one-size-fits-all containers. In an era without the constraints of physical shelf space and other bottlenecks of distribution, narrowly targeted goods and services can be as economically attractive as mainstream fare.*[4]

The long tail means that we can now serve previously under-served audiences. Prior to the Web it would have been extremely expensive to reach small audiences, but businesses like Amazon find that everything in their offerings is sampled once; perhaps not more than that, but at least once.

---

[4]Anderson, Chris, *The Long Tail* (New York: Hyperion, 2006).

The same applies to the engineering profession and its content. One may well ask who would want to hear an engineer talk about "plate efficiency" in a chemical engineering unit operation or listen to the details of how fiber optics work. Yet like Amazon.com and their infinite bookshelf, each of these videos would likely get at least one pairwise interaction because the topic resonates with someone. And that is precisely what engineering's long tail should do: match up interests and entries. This moves the mass media component of engineering outreach from an emphasis on big media hits - a television show or a *New York Times* article - to a world where, instead, 1,000 bloggers discuss in detail some aspect of science or engineering. What, then, are the details that make a social networking or wiki-style model work?

Wikis work!

Within academia the Wikipedia model gets little respect, yet for many subjects it works very well.[5] I use the site frequently and am often startled by the quality of information.[6] As of August 2010 Wikipedia has a bit over three million articles and is the third most popular site on the web behind Google and Facebook; the other top ten are all commercial.[7] So, Wikipedia's utility for millions of users has been settled. The interesting questions

---

[5]See The Chronicle of Higher Education's discussion among academics about Wikipedia at http://chronicle.com/live/2006/10/halavais/

[6]Errors, of course, occur, but that isn't unique to Wikipedia and new media. Recently I was reading John Hale's majestic *The Civilization of Europe in the Renaissance* (1994) - a 20th century masterpiece of history and a sterling example of "filter, filter more, then publish." On page 86 it announced that Francis I took over from his father Louis XII as King of France. Alas, Francis was a distant cousin. Unlike Wikipedia, this error will last for years and years.

[7]See *http://toolbar.netcraft.com/stats/topsites* for the most current statistics.

are why it works and how it can be used elsewhere.[8]

Four observations shed light on how the wiki model works -- whether it be text-, still-photo-, audio- or video-based. These observations are key to making an engineering new-media outreach project work.

- **Process, not product.** The key idea to keep in mind is that something like Wikipedia is not a product. Although the *-pedia* suffix makes one compare it to an encyclopedia, it is instead a process. A wiki doesn't work by collectivism but by continual and unending argumentation and emendation.
- **Centered on a debatable question.** A good wiki usually focuses on a question of the form "How does this work?" about an activity that its users want to engage in. For example, Flickr has a lively forum on HDR. Photographers make these High Dynamic Range images by combining three different exposures. This desire to do it oneself drives the forum.
- **Accommodate different levels of contribution.** Unlike a corporation, not all people who contribute to a project need to contribute equally. Some (many, in fact) do little, but a few do a lot. Why does this work here, but not in corporations and businesses? A car company, for example, must a) make cars and b) be a company. It takes a lot of work to be a company. Wikipedia, in contrast, doesn't need to be sure its employees show up. A company needs to ensure all workers are interchangeable and do the same amount of work; but Wikipedia contributors come and go. Return for a moment to the photos of Steuben County I mentioned earlier. As is typical of

[8]To test whether Wikipedia truly works, in October 2008 I created a short entry for a person worthy of inclusion: Professor Frances H. Arnold of the California Institute of Technology. I never edited the entry again, instead just letting it sit there. Others found it, added her correct birth date, inserted details of her work, and listed references.

a sharing site like Flickr or Wikipedia, the effort shows a skewed distribution: user *pawtrait04* contributed 1,547 photos, *kpmst7* 70, *danie.roman* 29, *Heron Hill Winery* 12, and *grockwell61* contributed 9 photos.

- **No experts.** Designating experts means no one writes an article. In a wiki, more people are likely to start a bad article than polish a good one. One must truly trust the "publish, then filter" model and let the filtering remove the most atrocious ones.

Still the writers need guidance. Jimmy Wales, one of Wikipedia's co-founders, notes:

> *Any company that thinks it's going to build a site by outsourcing all the work to its users completely misunderstands what it should be doing. Your job is to provide a structure for your users to collaborate, and that takes a lot of work.*

Not providing sufficient structure is the reason an experiment that *Wired* magazine carried out in "crowdsourced" journalism ended in failure.[9]

[9]Assignment Zero was an experiment in "pro-am" (professional/amateur) journalism, in which journalism is run by the public rather than the media. Assignment Zero was an attempt at journalism without strings — one might call it an audience-run newsroom. In the Assignment Zero project, stories were thought up, then chosen and researched by "citizen journalists," rather than designated by editors. The aim of this experiment was to promote social democracy — rather than the anarchy that one assumes would naturally result — and worked to employ a crowd model that allowed several contributors to shape a story. It failed.

4

How Engineers Should Talk to the Public

SO, WHAT should we tell the public? Are we experts that proclaim the correct answer to a scientific question? Are we primarily teachers whose goal is to create scientific literacy? Or, is there another role for us? For that matter, is our real battle for literacy, or is it against apathy?

As engineers entering the world of public communication, many obvious paths beckon us, but to my way of thinking they all culminate in dead ends – in a public that isn't more able to participate in democracy. I consider briefly these tempting paths before outlining a positive path. The unfruitful routes are: Presenting ourselves purely as experts, emphasizing research, using the word "technology" as a "catch-all" for engineering, and focusing on hard scientific literacy. These paths of explanation seem obvious, but on reflection are clearly the wrong approaches for talking to the public.

Being an expert

Being an expert is a legitimate and necessary role for an engineer, and it is especially natural for a professor. But ultimately it's very limiting. A major problem is that "expert mode" distances technology from the listener or viewer. When you use "expert mode," you say that science and engineering are something you cannot understand; you need my help. The gatekeeper role tends to turn off listeners. When on air I'm always called "Bill;" in fact, "professor" is never mentioned at all. Still, being viewed as an expert is a hard role to shake. Over the years I've gotten e-mails asking me for advice on building a concrete dome - part of someone's home improvement project - I've gotten questions about installing gas lighting, and even one about building a concrete submarine in the desert!

The cult of research: promises, promises

Another temptation that engineers face, especially as academics, is focusing on research and the research mission. Jacques Barzun, in his worthwhile *Science: The Glorious Entertainment*, calls it "The Cult of Research." "Research," he writes,

> *in other words, is no longer simply a vocation; it is an institution.*[1]

Barzun implies that institutional priorities overwhelm all others.

In the 1960s, academics formed a Faustian pact with entities like the National Science Foundation.[2] Essentially we said: Give us public funds for our *research* and we will return *technological* items of great value to the nation. Yet, the outcome of our research mission should not be to validate the technological products of research, but instead to emphasize the *human* product - the students and the

---

[1]Barzun, Jacques, *Science: The Glorious Entertainment* (New York: Harper & Row, 1964), p. 122.

[2]To be completely transparent about this: I *am* one of those academics - I'm a product and a producer; the U.S. Department of Energy paid for my M.S. and Ph.D., and my research as a newly-minted professor was funded by them.

knowledge created.[3,4] Think for a moment how impossible it is for a specific piece of research to return value in a regular and recurring way! At the core of research lies the strong possibility, even likelihood, of failure. Einstein famously said, "if we knew what we were doing, it wouldn't be called research, would it?" So, while we can make the argument of research's importance in the aggregate - over many cycles of research, our technological products improve the world - the nature of funding forces us to do this for *every* research proposal. This, in turn, results in overstatement and empty promises to the public. The degree to which we promise *direct* results from *specific* research is inversely proportional to how the public eventually assesses the value of the research *mission*. "We must not promise," wrote Lincoln, "what we ought not, lest we be called on to perform what we cannot." As a cautionary tale, consider the Human Genome Project.

The basis of this $3 billion project was to find the genetic roots of common diseases like cancer and Alzheimer's and then to generate treatments. Science writer Nicholas Wade

---

[3]Knowledge itself is different than the utility of the research results.

[4]This is best illustrated by the career of Bill Flygare at the University of Illinois. When I arrived at Illinois in 1984, the eminence of Flygare and his research still echoed through the halls, although he had died three years earlier at age 44. His colleague David Chandler reflected on Flygare's work two decades later: "The results of these experiments seemed important at the time. It was believed that knowledge of molecular dipole and quadrupole moments would significantly contribute to a good understanding of intermolecular forces. In current times, however, it is understood that intermolecular forces and their manifestations, especially in condensed phases, are more complicated than those numbers reveal. Further, experiment is no longer required for these quantities because theoretical quantum chemistry can now provide the information easily and reliably. In retrospect, therefore, the results of Bill's Zeeman effect measurements seem less important than the training provided to the students who helped make the measurements." *Biographical Memoirs,* volume 86, pp. 137-161 (National Academies Press, 2005).

concluded that "[t]en years after President Bill Clinton announced that the first draft of the human genome was complete, medicine has yet to see any large part of the promised benefits."[5] Even today, the old-fashioned family history does better than looking at the 101 genetic variants linked to heart disease.

So, to build a public outreach effort based on explaining research easily falls into the trap of reflecting the institutional priorities of the research mission – of getting that next grant, or proving to the funding agency the utility of the work.

Technology: a hazardous concept

If we use the word "technology," we defeat our outreach mission as engineers. On the surface what could be more odd? What more ridiculous claim could an engineer make? After all, "technology" nearly envelopes us: it fills our pockets with cell phones, it flies overhead as highly sophisticated jets, zooms by in the computer-controlled engine of a car, and flows through our veins as carefully designed drugs. In describing these items we use the word "technology" to lump together many separate, and even disparate, things.

For example, we might toss off a comment about how "railroad technology" has changed our world. Yet, what in the world do we mean by "railroad technology"? Well, we mean everything from drilling tunnels and laying tracks, to business organization, to telegraphy, to specially trained workers. Just to fully define the loaded phrase "railroad technology" takes a 700 plus-page book: *Empire Express:*

[5] Wade, Nicholas, "A Decade Later, Genetic Map Yields Few New Cures," *New York Times,* June 12, 2010.

Building the First Transcontinental Railroad.[6] We use "technology" as a catch-all which, while useful, lies at the heart of many myths about technology.

When we talk about technology as a single thing we give the sense that it has a mind and agenda all its own. Yet "technology" certainly isn't a thing; it is a human-made construct that reflects all aspects of humanity. Using the word technology buries the creativity of engineers: it camouflages the genius of engineers like George Stephenson or Isambard Kingdom Brunel.

Engineering creativity becomes buried because most people gauge the success of a technology by its degree of invisibility: the more concealed it is, the better! Think for a moment of heating your home in the 18th century. A typical house had an open hearth, which required *action* by the homeowner. It had to be filled with coal or wood, lit, and then constantly tended. Even so, much of this heat escaped through poorly insulated walls, prompting Theodore Roosevelt's wife to compare heating a home to "trying to heat a birdcage." Today, of course, we warm our well-insulated homes using a furnace hidden in the basement, pumping away with little assistance from its owner, and even less thought. Home heating has reached the pinnacle of technology: invisibility.

But this "out of sight, out of mind" means that the public no longer has visceral contact with technological objects, and so now believes myths about how this foreign "thing" called technology behaves. What we need to do in our outreach is make the work – the creativity, the inventiveness – of

---

6 Bain, David Howard *Empire Express: Building the First Transcontinental Railroad* (New York: Penguin, 1999).

engineers visible.

Hard scientific literacy

The final temptation is to engage in what's called "hard scientific literacy." What I mean by this is having a basic toolbox of skills - in mathematics, physics, chemistry, mechanics - that allows a person to delve into almost any technological area. Most readers of this book have such a toolbox. The goal of those who promote hard literacy is to create a public that is as capable as an engineer of making independent, scientific decisions. This has been the scientific literacy goal for the last thirty years or so. There is now a fair amount of evidence, though, that this effort has failed to penetrate the consciousness of the American public. In spite of all our efforts, by any reasonable measure we are a nation of scientific illiterates. If you just looked at the huge amount of work done to ingrain hard scientific literacy on a pragmatic cost/benefit basis, the effort would surely be abandoned.

Also, it isn't even clear that hard scientific literacy is desired. Morris Shamos, a physicist who's worked for thirty years to improve scientific literacy, reports:

> *To make matters worse, we keep insisting that public understanding of science means understanding some basic science rather than the technology that the public finds more palatable. All this despite the fact that ever since the Enlightenment, society has been sending back the message: give us the useful end products of science, as long as they cause us no real harm; but while we can relate to their technology, we don't require that we understand their underlying science.*[7]

So, if it isn't hard literacy we want - or can get - what do we

[7] Morris Shamos, *Myth of Scientific Literacy* (Rutgers University Press: Rutgers, New Jersey, 1997), p. 238.

aim for?

Hard literacy vs. awareness

We would like adults to understand how the scientific enterprise works in our particular political and economic climate. We want to encourage an appreciative public, one that at least understands how much needs to be spent on science and technology. The science and engineering community would be well served by a society that, while perhaps illiterate in science in the formal academic sense, is at least aware of what science is, of how it works, and of its horizons and limitations. You might call this approach "science awareness," rather than literacy.

The objectives of this approach are to help students, and society in general, feel more comfortable with new developments in science and technology. They need not understand the details but rather recognize the benefits and the possible risks of technology.

The real battle: technological determinism

The argument over hard literacy versus awareness distracts from the main problem. Our battle is not so much with literacy, as with technological determinism: the belief that technology shapes our lives with a ruthless logic all its own. In fact, which of us doesn't carry in their head an image of a great whirlwind of innovation that sweeps through our world, creating blessings and havoc? This view is only half true and, because of this, dangerous.

Its truth lies in the degree to which science affects our lives. Never before has such a complex web of technology permeated a culture. For sure, in every century some marvel has reshaped the world - the printing press, gunpowder, the cotton gin - but only in the twentieth century have these wonders united into a comprehensive system that seems poised to overtake us.

Technological determinism makes people passive and in doing so promotes a dangerous apathy. People become focused on how to adapt to technology, not on how to shape it. Thus, technological determinism removes a vital aspect of how we live from our public discourse, creating a pressing need for citizens who understand technological systems not just to grasp the impressive world of technology, but to exercise the civic duty of shaping those forces that shape our lives so intimately, deeply, and lastingly.

Lewis Mumford,[8] a pioneering historian of technology, pointed out that the products of engineering have meaning "only in relation to a human and social scheme of values." The key here is that the technical aspects of any technology cannot be construed apart from their social context. The values and world views, the intelligence and stupidity, the biases and vested interests of those who design a technology are embedded in the technology itself.

In my work, I look at the entire context of the things that surround us, which includes the people who make technology happen. In reaching out to the public, we must present the entire technical, social, political, economic, and cultural context of the things that surround us. This includes the innovators, inventors, engineers, entrepreneurs, and business people who make technology happen. And, more importantly, we must present this message in a way that

---

[8]In 1938, *Time* magazine featured Mumford on their cover. This is likely the one and only historian of technology to ever make the cover of a national news magazine!

resonates with the public.

The key aspect of how to do that comes from a G.K. Chesterton quote, which is posted in my office. Over the years I've collected quotations about writing and reaching readers, listeners, and viewers. I've laminated them and rotate them on and off a filing cabinet by my desk where I write. This is one of my favorites, and in fact is up there most of the time. Chesterton writes:

> *The only two things that can satisfy the soul are a person and a story; and even a story must be about a person.*

Look at very successful news shows like CBS's *60 Minutes* or NBC's *Dateline*. They always tell a story using, to my taste, too much suspense. They usually have a strong narrative, or at least a strong human interest angle. Look, also, for a moment at the scientific disciplines that are extremely popular: astronomy and evolution. They both had superb popularizers - Carl Sagan and Stephen J. Gould - but the public also sees them as situating us in our world. They answer questions like: Who are we? What is the purpose of life? The message to any technologist who wants to reach out is to place technology in context. As engineers we often neglect context, focusing instead on the mechanical details.

When I'm tempted to just explain how something works, I recall another quote that often rotates on and off my filing cabinet. Ambrose Bierce, in his 19th century *Devil's Dictionary,* essentially a list of literary barbs, defined "inventor" as,

> Inventor, *n. A person who makes an ingenious arrangement of wheels, levers and springs, and believes it civilization.*

This is what we must avoid when talking to the public.

To overcome the temptation to overhype my work, I often tell the story of an inventor or innovator who created some everyday object. I've talked about the invention of the microchip, Scotch tape, the Ping golf putter, and nylon. I'll use a story that reveals how technology is changing the listener's life or has dramatically changed our society. I've discussed the impact of the typewriter, the match, and how color film is embedded with cultural bias. Whenever possible, I like to link up technology with art, music, and especially literature. I've shared how J.R.R. Tolkien felt about technology and what his *Lord of the Rings* might mean for us today. I've delineated how the creative process of an engineer is closely linked to that of a painter. And, at times, I help listeners understand the news of the day. It isn't a mode I use often, but after the September 11th attacks, I tried to put technology and terrorism in perspective, and after the anthrax attacks I described anthrax and its toxicity in detail.

We can envision the general message we want to deliver, or rather the kinds of actions we want to see from the public. Let's take a more sophisticated look at the form of our message.

Framing

Journalism professors Matthew Nisbet and Dietram Scheufele have written engagingly and insightfully about how scientists and engineers should talk to the public.[9] Scientists and engineers, they note, tend to believe that facts will win out -- they call this the popular science model, which is a version of the hard scientific literacy I mentioned earlier. The authors cite ample evidence that this doesn't work. They review sixty years of

[9]Nisbet, Matthew C. and Dietram A. Scheufele, "The Future of Public Engagement," *The Scientist,* vol. 21, issue 10, p. 38, (2007).

research that suggests citizens prefer to rely on their social values. So, these authors argue for what they call "framing." In the abstract this means tailoring messages in ways that make them personally relevant and meaningful to different publics. Its best to look at some examples.

Let's start with the negative - the ways framing has been used against science. Greenpeace's idea of "Frankenfood" has been effective in opposing all manner of genetic modifications. This organization published a repulsive manipulated image of a frog's head on a rotting apple core. The image resonates deeply with some fear or social value that people have. What this means for technologists is that to reach out effectively is also to frame using what a group values. For example, when scientists talk to a group of people who think in primarily economic terms, that they should emphasize the economic relevance of science. An example might be embryonic stem cell research, pointing out how expanded government funding would make the U.S., or a particular state, more economically competitive. Nisbet and Scheufele praise E.O. Wilson's book *Creation: An Appeal to Save Life on Earth* for recasting environmental stewardship not only as a scientific matter, but also one of personal and moral duty, noting that this book has generated a discussion among a religious audience that might not otherwise pay attention to popular science books.

As a final example, consider the banner headline on the cover of a recent issue of Brown University's alumni magazine: "Could Today's Wonder Fiber Be the Next Asbestos?" This linking of ideas – nano and asbestos - echos something that buzzed across Europe. On the Continent the opponents of nano push it as the "the asbestos of tomorrow" or the “new asbestos.” This is, of course, framing in action. The public has placed asbestos in context - they have situated it in their political, cultural, and social landscape. This framing of nano ties it into their social judgment. European companies have responded with their own framing of “nano is nature” to try and tie into something else that citizens have already made a social judgment about.

Nisbet and Scheufele stress that they aren't talking about framing as “false spin;” rather they insist that the content be true. They argue, convincingly to me, that scientists engage in framing all the time. When writing a grant proposal, or a journal article, or providing expert testimony, scientists and engineers often emphasize certain technical details over others, with the goal of maximizing persuasion.

All that said, I find their approach too often focuses on only the value-laden scientific questions. Whenever I read their work, I cannot help but think of this quote from Justice Oliver Wendell Holmes from his great dissent in a case before the Supreme Court where he coined the phrase, “Great cases, like hard cases, make bad law.” He spells out the reasons:

> *For great cases are called great not by reason of their real importance in shaping the law of the future, but because of some accident of immediate overwhelming interest which appeals to the feelings and distorts the judgment.*[10]

He continues by citing the “hydraulic pressure” - his metaphor for growing peer pressure - applied by these interests. So, to my way of thinking we need to apply framing to less controversial and more everyday things.

---

[10]*Northern Securities Co. v.* U.S., 1904.

5

New Media in Action: Two Hypothetical Case Studies

TO ROUND out the more abstract ideas presented so far, I give here two concrete examples of how the engineering profession might use new media to achieve timeless goals. The first focuses on creating interest among teenagers about engineering, the second on educating the broader public about crucial issues with the power grid using "Citizen Science" methods. Where appropriate, I contrast the old and new media approaches.

Communicating engineering to young audiences

The U.S. faces a tremendous decrease in global competitiveness. As a measure, consider that the U.S. is now a net importer of high technology products (plus $54 billion in 1990 to a negative $50 billion in 2001). The seminal report *Rising Above the Gathering Storm*[1] highlighted the main element in reversing this trend: creating "a new generation of bright, well-trained scientists and engineers" who can "transform our future," noting that this must "begin in the 6th grade ...." The report mentions the need to "significantly enlarge the pipeline" of engineers, but as others have noted, this need is nuanced. It isn't the sheer number of new

[1]*Rising Above the Gathering Storm: Energizing and Employing America for a Brighter Economic Future* (Washington: The National Academies Press, 2007).

engineers that solves the problem but the type of engineer.[2] The NAE's report *The Engineer of 2020*[3] pinpointed the key issue:

> *Whatever other creative approaches are taken ....the essence of engineering - the iterative process of designing, building, and testing - should be taught from the earliest stages ....*

This means that we need to develop a cohort of pre-engineering students who have actually *done* engineering. In a field like engineering, nothing can replace "doing" because therein lies engineering's essence.[4] Thus, an important project for the engineering profession is to reach thirteen- to sixteen-year-olds who desire to create engineering projects in their time outside of class but lack both the information to make these projects successful and, although they don't know this, information on what really constitutes an engineering project.

Contrasting old and new media approaches

Oddly, the communication problem doesn't lie in the students' lack of interest in engineering as might be supposed. Research reported in the excellent NAE report *Changing the Conversation: Messages for Improving Public Understanding of Engineering*[5] found that tweens and teens very much resonated with the goals of engineering -- of creating a better, healthy, greener world.

---

[2]See, for example, V. Wadha, et al., "Where the Engineers Are," *Issues in Science and Technology*, Spring 2007.

[3]*The Engineer of 2020: Visions of Engineering in the New Century* (Washington: National Academies Press, 2004).

[4]For learning by doing, see Zhu, X., & Simon, H.A., "Learning Mathematics from Examples and by Doing," *Cognition and Instruction*, 4, 137-166, 1987.

[5]*Changing the Conversation: Messages for Improving Public Understanding of Engineering* (Washington: National Academies Press, 2008.)

Yet few could make the connection between their ideals and the work of an engineer. The report uncovered a critical step in engaging this age group: involve them in actual engineering. To do this, though, they need a community, information, interactivity, and role models that appeal to them. We could do this via old media or new media methods. Let's look at each.

The report *Changing the Conversation* bores in on creating a mass message. Using sophisticated polling methods they developed this positioning statement:

> *No profession unleashes the spirit of innovation like engineering. From research to real-world applications, engineers constantly discover how to improve our lives by creating bold new solutions that connect science to life in unexpected, forward-thinking ways. Few professions turn so many ideas into so many realities. Few have such a direct and positive effect on people's everyday lives. We are counting on engineers and their imaginations to help us meet the needs of the 21st century.*

A positioning statement lays the conceptual foundation for a communications campaign, but it is not usually shared directly with the public.

The old media approach - which has its merits - involves filtering of possible taglines. The very capable marketing firm hired to do this work developed taglines and then tested (filtered) how well they played with specific demographics focusing on the teen audience with the hope of enticing them to become engineers. Not surprisingly, no single message appealed to all groups, so they chose the "best" based on the teen sample and the marketers' considerable intuition. They chose a tagline for marketing: "Engineering because dreams need doing." By definition, the impact of this line is a compromise. The report goes on to suggest building a "public relations 'tool kit' to be used in advertising, press releases, [and] informational brochures." Could the same positioning statement be implemented with new media? Yes - and likely more effectively.

New media approach

Web 2.0 methods allow 13- to 16-year-olds to create content meaningful to them, instead of having to use an "educated guess" at a message. The content created will reflect their interests and style. Also, having a long Web 2.0 tail means that with the right new media message vehicle we can reach everyone - perhaps a particular experiment will be popular with only a few, but the cheapness of digital storage allows a description of this project to be kept up forever. Recall the pairwise matching mentioned in chapter three – that is, not a person speaking to a mass audience, but instead interacting via social media with two or three people at a time .

To engage 13- to 16-year-olds in actually doing engineering, one could create an organization that runs hundreds of after-school and summer camps for teenagers where they actual *do* engineering. The central new media piece would be a rich repository (a "long tail" in new media language) of step-by-step engineering project videos. This would be of a wiki format that allowed students themselves to add and emend projects; that is, to participate and thus bring the full power of Web 2.0 to the wiki. A key, though, would be seeding a wiki.

To seed the "long tail" of user-generated content, one needs photos, videos, audio, and text. Video would be generated by undergraduate students at engineering schools. Each of these schools would have what's called an

"Engineering Open House" in which - outside of class - the students build real, detailed engineering projects. The video blogging and wiki entries would document how their projects were done. Each could be rated and shared by the public. The key to making a successful long tail lies in uploading a huge amount of content fitted with social bookmarks that allow users to rate, comment, and forward video. A user of the site should be able to easily search for content, browse by subject or department, sort by rating, length, and so on, and rate and comment on videos. And the public should be able to upload their own "how-to" videos. Additionally, the videos should be easily downloadable to an iPod or other handheld device. The site should contain several RSS feeds:[6] one for all content, and feeds for specific subjects.

Why would such a site work?

- First, it focuses on the "How do you do it?" question so essential to making a successful wiki. In this case, tweens and teens doing science projects can see the details and contribute their own videos, rather than passively watching saccharine videos telling them how interesting engineering is.
- Second, it appeals to kids younger than the majority of participants in the videos. In his insightful book *Convergence Culture,*[7] Henry Jenkins notes that kids typically like to watch what people five to six years

---

[6]"RSS is a family of Web feed formats used to publish frequently updated works - such as blog entries, news headlines, audio, and video - in a standardized format. An RSS document (which is called a 'feed', 'web feed', or 'channel') includes full or summarized text, plus metadata such as publishing dates and authorship. Web feeds benefit publishers by letting them syndicate content quickly and automatically. They benefit readers who want to subscribe to timely updates from favored websites or to aggregate feeds from many sites into one place. RSS feeds can be read using software called an 'RSS reader', 'feed reader', or 'aggregator' which can be web-based or desktop-based." *Wikipedia.*

older than them are doing and model their behavior on that. This is part of the appeal of *American Idol*: half of its audience is composed of 13-year-olds who want to see an 18-year-old performing.

- Third, it offers long-term funding beyond the usual federal and foundation support through ad revenues because the audience of these videos is a prime demographic for advertisers. In the "old" media days, the restrictions on television and radio removed this opportunity.

Citizen engineering: Web 2.0 and the masses

In addition to reaching an audience of "future engineers," the engineering profession needs to tackle public literacy about engineering – to battle technological determinism rather than instill hard science literacy. The power grid offers a prime example for how knowledge creates a better citizen.

The public has great interest in solving the energy problems we face in the future but little knowledge of how to do so.[8] Currently 40% of our energy usage comes from electricity - power that is typically generated by coal, oil, and some nuclear. Clearly the United States will need to move toward alternative sources, and as that transformation occurs, the public will be faced with difficult choices. While they remain fascinated by these sources, the public rarely appreciates that these new energy technologies have an Achilles' heel: transmission.

As a recent issue of the *Economist* pointed out, "perhaps the greatest obstacle to the wider adoption of wind power is the

---

[7]Jenkins, Henry, *Convergence Culture: Where Old and New Media Collide* (New York: New York University Press, 2006).

[8]Mckeown, Rosalyn, "Energy Myth Two - The Public Is Well-informed about Energy," *Energy and American Society – Thirteen Myths,* edited by Benjamin K. Sovacool and Marilyn A. Brown (Springer Netherlands, 2007).

need to overhaul the grid to accommodate it."[9] So, as the United States moves toward alternative energy sources, for its citizens to be effective -- as voters or in applying knowledge in their own communities -- they need to understand the electrical grid.

Many electrical and power engineers feel the public underestimates the difficulty with which renewables can be added to the electrical grid. In the grid's nuances and peculiarities lies a major hurdle to using and incorporating non-fossil-based alternatives into our nation's energy mix. The public rarely thinks of the grid, yet it is the nervous system of our nation's energy infrastructure. We often concentrate on that system's "heart" (the generation of power by coal, oil, hydro, or nuclear) but rarely think of its transmission. To facilitate that understanding and to create literate citizens, the project described below makes citizens an active part of monitoring and developing the new grid. The long-term goal is to develop smart meters for an intelligent grid - a grid designed to be more responsive to changes in load and designed to give feedback to consumers. One way to achieve literacy about the grid is to use a "Citizen Science" approach enhanced by Web 2.0 techniques.

Citizen science

The development of social networking tools has given new impetus to Citizen Science, which Wikipedia aptly defines as:

> *.... a term used for projects or ongoing programs of scientific work in which individual volunteers or networks of volunteers, many of whom may have no specific scientific training, perform or manage research-related tasks such as observation,*

[9]"Wind of Change," *The Economist Technology Quarterly,* December 6, 2008, p. 22-25.

measurement or computation.

Citizen Science projects have engaged the public to classify over a million star clusters, to collect data on ecosystems, and to help researchers better understand birds and their habits.[10]

Web 2.0 social tools, then, offer the promise of vastly expanding citizen science projects and of increasing their efficacy. Bruce V. Lewenstein, Professor of Science Communication at Cornell University, notes two additional benefits of Citizen Science to the engineering profession: 1) the engagement of non-scientists in true decision-making about policy issues that have technical or scientific components; and 2) the engagement of researchers in the democratic and policy processes.[11]

In many ways engineering lends itself better to this approach than does science because it is a process-oriented activity with a teleological goal of producing something, whereas science is focused on discoveries about nature. Consider a project that would inform citizens about current issues, problems, and research on the power grid.

Power engineers need to know, to meet the U.S. energy needs of the 21st century, how new technologies affect the grid. They would like to know, for example, what happens to the grid if everyone installs compact fluorescent bulbs. Or, what if the sales of hybrid cars skyrocket? How would cars plug into the grid at night? In their studies of "load

---

[10]See *www.galaxyzoo.com* for details on stars, for ecosystems see C.B. Cooper, et al. "Citizen Science as a Tool for Conservation in Residential Ecosystem" *Ecology and Society* 12(2) issue 11 (2007); for studies with birds refer to the work of the Cornell Lab of Ornithology.

[11]Lewenstein, Bruce V. "What does citizen science accomplish?" Paper read at CNRS colloquium, 8 June 2004, in Paris, France.

modeling" - a very important topic in the field now - they are concerned with "power electronics" in the home. A home solar system, for example, contains these power electronics. These devices can really mess with the grid.[12] To understand the grid's behavior more deeply and to monitor it more closely, engineers need many independent observations; exactly where a citizen engineering project excels.

Citizen engineering methods make use of the thousands of eyes and brains of their participants to gather and, in some cases, to analyze data that stretch across a large distances. For example, Cornell's Lab of Ornithology uses Citizen Scientists to track birds across the U.S. In the power grid example, a frequency meter, power meters, and other devices would be located in the home of every participant.

The project would use a new technology, called FNET (frequency monitoring network technology), developed at Virginia Tech by YiLu Liu, a professor of electrical and computer engineering and an expert on the electrical grid.[13] She and her team have developed a small box - called a Frequency Disturbance Recorder - to measure changes in frequency on the grid.

The simplicity of the technology from a user's point of view is rather astounding. There are no installation costs; the

[12]A simple definition for power electronics would be "the control of 'raw' input electrical power through electronic means to meet load requirements." Power electronics is interdisciplinary and is at the confluence of three fundamental technical areas: power, electronics, and control.

[13]S.-J. S. Tsai, J. Zuo, Y Zhang and Y. Lui, "Frequency Visualization in Large Electric Power Systems," Power Engineering Society General Meeting, 2005 (IEEE), Issue 12-16, June 2005, p. 1467-1473; L. Nystrom, "Energy Grid," Virginia Tech Research, Summer 2006, p. 1-5.

user just plugs a unit into a standard electrical outlet. What is the value in knowing the frequency at many points on the grid? The grid generates power at a specific frequency of 60 cycles-per-second. If any part of the grid deviates by as much as 1/20 cycles per second, trouble develops. If it drops to 59 cycles per second then havoc, such as the blackout of 2003, results. The frequency, then, is akin to a human pulse: its measurement and value tells us something about the health of the grid. With the fifty or so devices that Professor Liu has employed across the Eastern Interconnect - the grid that powers the mid-west, eastern seaboard, and parts of Canada – she has detected, earlier than anyone else, disturbances in the grid. The goal of a power grid project would be to deploy thousands more of these devices and to get fine-scale information about the current health of the grid and about the grid's behavior as we add renewables and power electronics. Currently there are 50 or so meters out there. With 2,000 or more researchers, we could truly understand the grid at a very local level, thus preventing disturbances nationwide and providing the essential data for adding renewables. Recall that the grid is both highly local and interdependent. The blackout of August 2003, for example, occurred because a small northern Ohio power company failed to trim a tree along a power line.

To incorporate these local frequency and power meters into a true citizen engineering project, we would need to develop the proper cognitive tools (e.g., a wiki) so that each participant would be able to see, share, discuss, and enhance their own observations. One can picture a visualization software package for "citizen engineers" that shows a large amount of information in a single computer-generated image - images that are useful, even indispensable, to monitor the electrical grid. These would allow a citizen to track flows of electricity in their own community and to see how they are linked, and thus interdependent with much of the rest of the nation. In short, visualization allows a member of the public to comprehend the grid by lifting the truly significant events out of the background noise. The power meters could also be used to locate "energy vampires"[14]- in a home. The wiki will allow users to share thoughts and offer suggestions on how to improve energy efficiency in their houses. An outcome of this enhanced interest would be a new cache of public knowledge about the power grid: its prowess, its promise, and its limitations.

[14]Those electronic devices that silently suck away energy even when they appear to be turned off.

6

Conclusion: Creating Zing!

EVEN THOUGH I noted at the outset that this short book would not teach you the latest social networking tools, you may feel just a little at sea, wondering, "What exactly should I do now?" Although my intent lay in changing engineers' viewpoint about media, you deserve a more concrete, less elusive answer.

As with any outreach to the public, we want to both raise awareness and increase engagement, but new media offers us an additional distinction: The ability to create a social movement and to effect social change. The snappy word "zing" captures it all. It is a notion that marries the oldest concept of outreach - of engaging intellectually and emotionally – to social media's ability to involve and engage the public by creating direct collaborations with them. With new media, we no longer measure engagement simply by audience numbers, but by creating "deep use" among those we want to reach. We want them to create media "as contributors, amplifiers, sharers, raters, commentators, distributors, re-mixers -- forming their own organic network."[1]

Thus the engineering profession needs to create new

[1]Clark, Jessica and Sue Schardt, *Spreading the Zing: Reimagining Public Media through the Makers Quest 2.0* (AIR Perspective 2010).

media projects that become "loved" by their users, loyal users who sustain, nurture, and guide them. Clay Shirky once argued that Wikipedia thrives and survives because of . . . love! Indeed, create an irrelevant page on Wikipedia - perhaps a PR piece about your firm - and "MrOllie" and his gang will quickly delete it and send you a curt note on "Conflicts of Interest." By some estimates, if MrOllie and gang stopped their work, Wikipedia would end within a month. Unpaid, they do this work because of their admiration for the concept of Wikipedia.

My hope is that this book lays down a foundation for engineers to create a grand new media project that becomes a powerful method for public outreach. I hope they use this book to design such a project so that it has a chance of working from day one. I do not know the form this should take, I have not the wit to design such a thing alone; but in this interconnected, wiki-based age, many minds should be able to create it. I do know the elements of what we want this entity to do and how it should work:

> **Catalyze user participation.** It must focus on some kind of "how" question; something that users can judge and comment on. If the project becomes simply a repository, then it will fail. It must create love and in turn have utility.
>
> **Create the right kind of engineer and foster in the public the proper notion of engineering.** This means that the project must reflect the complexity of engineering, rather than the reductive approach commonly used in our engineering schools. By complexity I mean that an engineering solution reflects an interdisciplinary approach that uses both technical depth and non-technical breadth. As one commentator put it, our goal should be "to be able to adapt and apply

technology that is human-centered, desirable, feasible, viable, sustainable, usable and manageable."[2]

Use the low cost of failure to succeed. Creating such a new media project will take enormous creativity, great skill, and a large amount of luck. The latter implies that it will come about because many projects are tried, and only a few succeed. Recall that a hallmark of new media is the low cost of failure.

Embed the notion of outreach in the "DNA" of every new engineer. In the day and age where the line between personal and public communication has blurred, and where citizen journalism might well dominate, we need to have every engineering graduate versed in new media, and in love with the idea of reaching the public.

---

[2]Craig, Kevin, "Complexity Demands a New Engineering Mindset," *Design News,* June 22, 2010.

Further Reading

As a media practitioner and not a theorist, I have made no original contributions in this book to understanding new media. Instead I have applied the insights and even used the examples from the books listed below. I found them extremely helpful in getting my bearings in the landscape of new media, but bear in mind that the best way to learn new media is to be a regular user of it. Nothing teaches in this area like participation.

Anderson, Chris, *The Long Tail: Why the Future of Business is Selling Less of More* (New York: Hyperion, 2006), 238 pp.

Anderson wrote this essential text in an approachable, even breezy style - as one would expect from the editor of *Wired* magazine. He lays out clearly how the digital distribution of goods changes fundamental ideas about commerce, although he overstates a bit the degree to which the "future of business is selling less of more." Still it has great utility in showing how and why engineering communicators can now aggregate audiences previous unavailable to them.

Heath, Chip, and Dan Heath, *Made to Stick: Why Some Ideas Survive and Others Die* (New York: Random House, 2007), 294 pp.

I have not explicitly used this text in the book, yet it influenced deeply the examples I chose. It is *the* book I recommend to all who want to communicate engineering more clearly and effectively, even though you will not find one iota of science or engineering in it. The Heath brothers delineate clearly why most communication fails and give clear guidance on how to make it succeed. I use their approach so often that I made a four-page outline of *Made to Stick's* essential ideas. You can find a copy of my outline at *www.engineerguy.com*.

Jenkins, Henry, *Convergence Culture: Where Old and New Media Collide* (New York: New York University Press, 2006), 308 pp.

Written by an academic, this book delves deeply into the interaction of new and old media. It reminds that the impact of new media is not just Facebook and Twitter, but the intersection of the internet with traditional forms like movies and television. His chapter on *American Idol* will give you deep insights into why such a show works; his analysis of *The Matrix* highlights the impact of social media on a film franchise. An eye-opening book; although, as one would expect from case studies, it is very detailed.

Shirky, Clay, *Here Comes Everybody: The Power of Organizing Without Organizations* (New York: Penguin Press, 2008), 326 pp.

Shirky's thesis that we will all work together in different ways because of social media need not detract a reader from his insightful analysis of what makes new media tick. In this regard he is unparalleled. Read this and the *Long Tail* and you'll have all the new media "theory" you need to pass safely through the Web 2.0 portal.

About the Author

BILL HAMMACK'S work has been recognized by an extraordinarily broad range of scientific, engineering, and journalistic professional societies. From his engineering peers he's been recognized with the ASME's *Church Medal*, IEEE's *Distinguished Literary Contributions Award*, ASEE's *President's Medal*, and the AIChE's *Service to Society Award*. From journalists he has won the trifecta of the top science/engineering journalism awards: The National Association of Science Writer's coveted *Society in Society Award*; the American Chemical Society's *Grady-Stack Medal* - an award previously won by Isaac Asimov and Don Herbert (Mr. Science) - and the American Institute of Physics' *Science Writing Award* -- all typically given to journalists. *Make Magazine* described Bill as a "brilliant science-and-technology documentarian" noting that his recent video work "should be held up as models of how to present complex technical information visually."

Hammack, a Professor of Chemical & Biomolecular Engineering at the University of Illinois – Urbana, is a leader in using mass media to communicate engineering to the public. Pioneering a new role for an engineering professor, he created a remarkable public radio series called "Engineering & Life," in which he shared with the public the wonder of engineering, while also emphasizing the responsibilities associated with technological change. His hundreds of radio pieces have been heard on public radio's premier business program *Marketplace*, which has an audience of six million, and around the globe on Radio National Australia's *Science Show*.

In 2005-06, he broadened his "audience" to include senior government policymakers. He served a year as a Senior Science Adviser at the U.S. Department of State. At the U.S. Department of State, Hammack served as an energy adviser for the Six-Party Talks to denuclearize the Korean Peninsula, helping to develop a framework for U.S. negotiations. Additionally, he served in the Department of State's Bureau of International Security and Nonproliferation, representing the U.S. in successful talks with Vietnam to remove highly enriched uranium, which can be used to make a small nuclear bomb. Through his pioneering work, he is creating technologically literate citizens and government officials who will have a huge impact on the health of our democracy, our national economic productivity, and foreign policy.

Why Engineers Need to Grow a Long Tail

A Note on the Type

The text for this book was set in Bembo. In 1929, Stanley Morrison created, for Monotype, a 20th-century revival of an old-style serif or humanist typeface first cut by Francesco Griffo around 1495. Originally trained as a goldsmith, Griffo's typeface departed from the slavish dependence on pen-drawn characters. Griffo's precision skills, acquired from engraving steel, allowed him to refine the type far beyond that of a pen. Bembo - named after its first use in the book *Journey to Mount Aetna* written by the young Italian humanist poet Pietro Bembo - features nearly all the characteristics that define old-style humanist designs; for example, a minimal variation in thick and thin stroke weight and angled top serifs on lower-case letters. It proved very popular with British publishers. In the 1930s, book designers chose it frequently, making it a dominant letter form.

Chapter headings were set in Perpetua, a typeface designed by British sculptor, typeface designer, stonecutter, and printmaker Eric Gill (1882–1940). It is a transitional typeface because of its high stroke contrast and bracketed serifs. Gill began work on Perpetua in 1925 at the request of Stanley Morison, typographical adviser to Monotype. Perpetua's first use, appropriately, was in a limited edition of a new translation by Walter H. Shewring of *The Passion of Perpetua and Felicity* (1929). It appeared most recently on the covers of the Artemis Fowl book series and in Barack Obama's 2008 campaign logo.

As appropriate for a book on new media, much of the information above came from *Wikipedia*.

Made in the USA
Charleston, SC
19 December 2010